WP

Wakefield Libraries
& Information Services

This book should be returned by the last date stamped above. You may renew the loan personally, by post or telephone for a further period if the book is not required by another reader.

JOHN DEERE
TWO-CYLINDER
TRACTORS

JOHN DEERE
TWO-CYLINDER
TRACTORS

MICHAEL WILLIAMS

B **Bounty**
Books

First published by Farming Press Books and Videos in 1993

This edition published in 2007 by Bounty Books,
a division of Octopus Publishing Group Limited,
2-4 Heron Quays, London E14 4JP

ISBN-13: 978-0-753715-05-5
ISBN-10: 0-7537-1505-8

A CIP catalogue record for this book is available
from the British Library

Printed and bound in China

Contents

Acknowledgements

THIS book would not have been possible without the co-opera-
tion of John Deere two-cylinder tractor owners in America,
Canada and England. I would particularly like to thank all those
owners who allowed me to photograph their tractors, many of
them going to considerable trouble to make the tractors available
for the camera.

I also wish to thank Deere & Co for providing some of the
photographs published in this book, and for making data on serial
numbers and production years available.

The photographs on pages 13, 16, 61, 70 (below), 71 (above and
below), 72 (above and below), 73 were taken by David Williams.
The photograph of a Waterloo Boy tractor on page 12 is from the
Smithsonian Institution. John Deere supplied the photographs on
pages 1, 3, 7, 47, 83 (below), 86; and the pictures of the Froelich
tractor on pages 4, 5, and 6 are from the Two-Cylinder Club. The
rest of the photographs are from the author's collection, most of
them taken especially for this book.

Additional Information

THE principal source of information about John Deere tractor history is the Two-Cylinder Club, an American-based organisation with more than 25,000 members worldwide. Benefits of membership include six issues a year of the *Two-Cylinder magazine*, and members may also access the club's research facilities to obtain specific information.

Membership inquiries should be sent to the Two-Cylinder Club, Grundy Centre, IA 50638, USA.

Additional information is also available from a number of recently published books. Volume one of *John Deere Tractors and Equipment*, which deals with the period from 1837 to 1959, features detailed information and a large number of illustrations of equipment as well as tractors. The authors are Don Macmillan and Russell Jones, and the publisher is the American Society of Agricultural Engineers. A second volume covers the period from 1959.

The ASAE also publishes *John Deere Tractors 1918–1987*, a soft-cover book containing information including production data, plus line drawings and black and white photographs.

How Johnny Popper Replaced the Horse was published by Deere & Co in 1988. The book is illustrated by a large number of colour photographs and covers the development of the two-cylinder tractor range, plus an extra chapter on John Deere construction equipment. The authors are Don Huber and Ralph Hughes of the Deere Company advertising department.

John Deere Tractors – Big Green Machines in Review is written by Henry Rasmussen and published by Motorbooks International. It has a large number of colour photographs showing both restored and 'working' tractors in the United States.

To Shiela and Don Morden

1

John Deere and John Froelich

JOHN DEERE is one of the great names in the farm equipment business. Deere & Co, the company which makes the John Deere range, is one of the world's biggest and most successful manufacturers of tractors and machinery and is also one of the oldest with a history dating back to a small blacksmith's shop in Grand Detour, Illinois in the 1830s.

The blacksmith's name was John Deere. He was born in 1804 in Rutland, Vermont where his parents had settled after emigrating from England. By the age of 19 John was working for a local blacksmith, and he later started his own business making simple hand tools for farmers in the area.

John Deere

In 1836 John Deere decided to give up his business in Vermont and move further west to Illinois, where he became the local blacksmith in Grand Detour, which was then a small village. It was a time when crop production was expanding rapidly on the fertile soils of Illinois and throughout much of the Midwest, creating a strong demand for farm equipment.

The product which brought John Deere his first big success was a new plough (in American spelling, plow) which he developed in 1837. Many of the farmers who moved into the area found that the soil could be sticky and difficult to cultivate with conventional ploughs. John Deere solved the problem by designing a different type of plough with an effective self-cleaning action. The shape of the mouldboard was improved, but the real breakthrough was using high grade steel which was polished to allow the soil to slide off more easily. The steel was also harder and more wear resistant than the iron mouldboards used previously.

The idea of using steel instead of iron came when John Deere noticed a broken saw blade while visiting the local sawmill. The blade was made of high quality steel and was polished, and he decided to take it back to his workshop where it was used to make an experimental mouldboard and share which were tested on a nearby farm. The plough performed well, with the shiny steel mouldboard providing an effective non-stick surface.

Orders began to arrive from local farmers who had heard about the improved ploughs, and John Deere had to expand his production to meet the growing demand. One of the earliest John Deere ploughs, made in about 1838, is preserved at the Smithsonian Institution in Washington DC.

Sales continued to increase as the reputation of the new steel ploughs spread further afield, and in 1848 John Deere decided to move again, this time to Moline on the borders of Illinois and Iowa.

Moline was a better distribution centre for the expanding business, and the small factory which John Deere built there in 1848 was extended in the following year as the demand continued to grow. More factory extensions were needed as production reached 13,000 ploughs a year by the mid 1850s and continued increasing to more than 40,000 implements of all kinds in 1869, the year after Deere and Co had been established with John and his son Charles among the principal shareholders.

One of the reasons for the company's success was John Deere's concern for quality. This was a priority at every stage in his business career, from his work as a blacksmith in Vermont to running one of America's biggest implement factories in Moline, and the company had established an excellent reputation by the time John Deere was ready to hand over the control of the company to his son.

John Deere died in 1886 at the age of 82. By this time the company he had started 50 years earlier was selling well over 100,000 horse drawn implements a year and was still growing.

Three years after John Deere's death another American named John Charter of Sterling, Illinois built what was probably the world's first tractor, using a single cylinder petrol or gasoline engine mounted on a steam engine chassis. The tractor was used to power a threshing machine on a farm near Madison, South Dakota where it performed well enough to attract orders for five or six more tractors based on the same design.

In spite of its modest success the Charter tractor aroused only local interest at the time, but it was the beginning of a new age in mechanised farming which was to have important consequences for the future of the John Deere company.

John Froelich

More tractors appeared in 1892. The J I Case Threshing Machine Co of Racine, Wisconsin built an experimental tractor, but then abandoned the idea to concentrate on steam traction engines, and the Dissinger brothers of Wrightsville, Pennsylvania built their first tractor in the same year.

The third of the 1892 tractor pioneers was John Froelich of Froelich, Iowa, a village which had been named after his father. John Froelich was born in 1849, and in the late 1880s he owned a steam engine and a threshing machine which worked on farms in Iowa and South Dakota. He decided to expand his threshing business and in 1892 he built a tractor, mounting a single cylinder engine on a chassis he designed and built with the help of a local blacksmith. The tractor was probably the first to be equipped with both forward and reverse gears, and it was taken to South Dakota where it was said to have threshed 72,000 bushels in 52 days.

The gasoline engine chosen for the Froelich tractor was made by

The 1892 Froelich tractor

the Van Duzen company of Cincinnati. The single cylinder had a 14in bore and 14in stroke which, as the Two-Cylinder Club points out in an introduction to their Expo 111 event celebrating the tractor's centenary, is a capacity of 2155 cubic inches or 35.5 litres. The cylinder was vertically mounted and the power output was 20hp.

A year later, in 1893, John Froelich and a group of businessmen from Waterloo, Iowa formed a company to develop his tractor commercially. The company, which was called the Waterloo Gasoline Traction Engine Co, built and sold at least two tractors in 1893, but both were returned to the company because the customers were not satisfied with their performance.

After this setback they decided to concentrate on stationary engines, for which there was a growing demand. The company was reorganised in 1895 when the name was changed to the Waterloo Gasoline Engine Co, and John Froelich left in the same year.

Close-up of the Froelich tractor transmission

One tractor was built in 1896 followed by another in 1897, but apart from these the company relied on stationary engine production. The Waterloo Boy name first appeared on a new range of engines introduced in 1906, and this became an increasingly familiar trade name for the next 18 years.

Meanwhile the demand for tractors was expanding and more companies were moving into the market. Deere and the Waterloo Co both remained on the sidelines for several years, but both were attracted to what was clearly a growth market.

Both companies had an obvious interest in the development of the tractor market. Deere and Co was already supplying gang ploughs for use with the tractors which were beginning to replace horses and steam engines. The fact that International Harvester, a leading machinery company, was already achieving considerable success as a tractor manufacturer may also have encouraged the Deere directors to take the tractor market seriously. This view would have been reinforced when another prominent farm equipment company, J I Case, returned to the tractor market in 1911, almost 20 years after their first venture.

The Froelich tractor in a photograph taken in South Dakota on September 30th 1892

This replica of the Froelich tractor was built to celebrate the Deere & Co centenary in 1937

The decision to begin development work on a John Deere tractor was taken in 1912, and this was followed by a series of experimental prototypes including a three-wheel drive model designed by Joseph Dain. His tractor was known as the All-Wheel Drive, and a preliminary production batch of 100 was authorised in 1918 after four years of development work and testing to improve the design.

The Waterloo Gasoline Engine Co was interested in the tractor market as an additional outlet for their engines. The company returned to tractor production in 1912 with the 25hp Waterloo Boy Standard model powered by a four-cylinder transverse engine. It was followed a year later by a half-track version with short tracks instead of driving wheels. The Standard tractor was not a big commercial success, but the Waterloo Boy Model R which arrived on the market in 1914 was more popular and further success was achieved in 1917 when the company announced an improved version known as the Model N.

It was the Model N which took John Deere into the tractor market after Deere and Co paid $2,350,000 to take over the Waterloo Gasoline Engine Co in 1918, and the Model N remained in production under the Waterloo Boy name until the John Deere Model D arrived in 1924.

2

Waterloo Boy Models R and N

THE Waterloo Boy Models L and R arrived on the market in the early stages of a massive increase in the American tractor market. Tractor production in the United States was estimated at about 15,000 in 1914, but five years later the total had reached more than 160,000 and was still growing.

When the 1912 Standard tractor was developed the Waterloo Gasoline Engine Co had chosen an advanced design with a four-cylinder engine and automobile type steering. The four-cylinder engine was replaced by a two-cylinder design when the new L and R models arrived, and this was the beginning of the engine design which remained a distinctive feature of John Deere tractors until 1960.

Engine detail of a 1918 Model R

Above
A 1924 Model N owned by the Budd family of Ingersoll, Ontario

Top left
Chain steering on a 1918 Model R owned by Tony Ridgway of Ohio

Bottom left
Worm and sector steering on a Model N

Exactly why the designers switched back from four cylinders to two is not clear, but lower production costs and greater simplicity were probably important reasons. An engine with two cylinders costs less to build than a four cylinder design, and this helped to keep the price of the tractor more competitive. The advantages of a more simple two-cylinder design include fewer parts and less risk of a breakdown, and it also helped to make the engines easier to maintain and adjust for customers who were more familiar with horses than with tractors.

When production of the Model L and R tractors started in 1914 the worm-and-sector automotive steering on previous Waterloo Boy tractors was replaced by the chain operated mechanism which had been standard equipment on steam traction engines for many years. Chain steering was becoming outdated, but it continued until the end of the Model R production run in 1919 and was also used on the Model N until automotive steering was re-introduced by Deere in 1920 on an improved version of the Waterloo Boy.

The cylinder dimensions of the horizontal engine designed for the Model R were 5.5in bore and a 7in stroke, but the bore was increased to 6in during 1915. A further 0.5in was added in 1917 to take the Model R bore to 6.5in, and this was the version used when Model N production started in the same year. Another development was the addition of a removable cylinder head, a feature introduced during 1916.

As well as modifying the steering and the engine, the Waterloo Co frequently made other design changes to the Model R, with four different versions being built during the first nine months of production. The most significant developments included a two-speed gearbox for the Model N instead of the single speed of the previous models, and the original bolted construction of the main frame was replaced by a riveted design during 1922 after the Deere takeover. Another improvement was using roller bearings on the Model N instead of the plain bearings on earlier models.

1918 Model N in the Smithsonian Institution collection

Right
Driver's eye view of a Model N

Belt driven cooling fan on a later Model N

In spite of the improved bearings, lubrication was still a demanding chore for the Model N driver. The Manual of Instruction has a list of 18 oil or grease points which needed attention twice each day, ranging from two drops of machine oil on the valve guides to 30 drops on the pulley shaft bearings. A further nine points required lubrication once per day, the centre pin on the automotive type front axle and the worm gear of the steering drum both needed attention twice each week, and the rear axle cannon bearings and differential bearings needed oiling once per week. In addition to all this there were separate lubrication instructions for the magneto.

In spite of its basically conventional design, the Waterloo Boy Model N had an important part to play in the history of farm tractor development, achieving fame in three different ways. It brought Deere and Co into the tractor industry, and it was the first tractor to complete a Nebraska test (see appendix A).

The Model N also influenced tractor history by helping to bring Harry Ferguson into the tractor industry. This happened during the 1914–18 war when Britain was threatened with starvation because of the success of German U-boats in cutting off imported

14

The Budd family's Model N ready for starting

Overleaf
1918 Model N at the Great Dorset Steam Fair in England

food supplies. The British Government realised that tractor power offered the fastest way to increase food production from British farms, and large numbers of tractors were imported from the United States.

These included about 4000 Waterloo Boy tractors, mainly Model Ns, which were marketed in Britain as the Overtime. At that time Ferguson was running a successful garage business in Belfast, and the wartime increase in sales persuaded him to add tractors to his other interests. The tractor he chose was the Model N Waterloo Boy or Overtime, and he became the distributor for what is now Northern Ireland. The experience he gained demonstrating and selling Overtimes was the start of an interest in tractor design which eventually resulted in the development of the Ferguson System of three-point linkage with hydraulic controls.

Model N production continued until 1924 while the Deere engineering staff were working on a replacement tractor. The production total for the N eventually reached 20,534, compared with more than 8000 of various versions of the R and just 29 of the Waterloo Boy L and LA tractors.

Final drive of a 1918
Model R . . .

. . . and of a late
Model N

3

John Deere D, C and GP

**1925 Spoker D
no 32015 in the Budd
family collection at
Ingersoll, Ontario**

ALTHOUGH the Model N established Deere in the tractor market, it remained basically a Waterloo design with some Deere improvements, and it continued to sell under the Waterloo Boy name.

The N was replaced in 1923 by the Model D, the first tractor designed by Deere and sold in commercial numbers under the John Deere name. It was a much more up-to-date design than the R and N tractors had been, and it was developed for a more competitive market. It became one of the great success stories of the industry with a production run which continued until 1953.

Development work to produce a replacement for the Model N was already underway at the Waterloo factory when John Deere took the company over. The engineers had been working on the design for a tractor with fully enclosed transmission, a project which was continued under the new management. Deere engineers had also been actively involved in several development projects including the Dain tractor.

In spite of all the development activity in the period before 1920, there was no attempt to rush a new John Deere tractor on to the market. The priority was on careful development and testing, just as it was during the Dain tractor development, and the first Model Ds did not leave the factory until the summer of 1923.

Serial plate for 32015

Right
Spoked flywheel of 1924 D no 31137 owned by Chris Cobler of Ottumwa, Iowa

Original brass Schebler carburettor and primer on a 1926 D owned by Carl Roney, Elmira, Ontario

The first production version of the Model D was equipped with a 26in diameter spoked flywheel which was keyed on to the crankshaft. After fewer than 900 tractors had been built to this design the flywheel diameter was reduced to 24in, which helped to make the two-cylinder engine more responsive. Another change came at the beginning of 1926 when a solid flywheel replaced the spoked version, and in the following year this was modified again when the crankshaft and flywheel were splined.

Model D tractors with a spoked flywheel, known to enthusiasts as 'spoker Ds', reached a production total of about 5755, and the survivors are highly prized by collectors.

My namesake, Michael Williams of Clinton, Iowa, owns this beautifully restored 1936 Model D no 126185

The diameter of the spoked flywheel had been reduced to allow extra space for the straight steering rod, which was located on the left hand side of the tractor, and at the same time the original steering rod was changed to a jointed design which survived until worm-and-gear steering was introduced in 1931.

The engine design was inherited from the Waterloo Boy tractors, but the rated rpm went up from 750 to 800 when the D was introduced, with a further increase to 900 as part of a batch of modifications in 1931. The 6.5in cylinder bore for the first five years of D production was uprated slightly to 6.75in from 1928 onwards. The series of engine modifications increased the power output recorded at Nebraska from 30.40hp in the 1924 test to 41.59hp in 1935.

A three-speed gearbox was added to the specification in 1935, and the D was included in the styling changes introduced in 1939.

Text continued on page 29

Left
**Flywheel of 126185 and
. . .**

Right
**. . . the rice lugs on the
rear wheels**

Overleaf
**A 1938 Model D on
rubbers, another
tractor from the Budd
collection**

Streeter D no 191624 made in 1953 and owned by the Cobler family of Ottumwa, Iowa

1928 Model C no 200109 owned by Bruce Keller, Brillion, Wisconsin. It was shown at the 1993 Two-Cylinder Center Grand Opening

Engine of 200109

Demand for the Model D increased quickly and sales passed the 10,000 a year level in 1928. Meanwhile the company's engineers were working on a new model for row-crop work. International Harvester had already scored a major success with their Farmall tractor designed by Bert R Benjamin. At a time when the tractor market was dominated by Henry Ford's price cutting policy for the Fordson, Benjamin had realised that a different type of tractor was needed to mechanise row-crop farming, and the Farmall was the result.

Front view of 200109 showing the sharply arched front axle

Deere's first production row-crop tractor was the Model C. It had an arched front axle to straddle the rows, with plenty of ground clearance to accommodate mid-mounted equipment. It was also designed for either standard or tricycle front wheels, and Don Macmillan and Russell Jones in *John Deere Tractors and Equipment* record that one of five pre-production prototypes built in 1926 was a wide tread version.

The first Model C tractors left the Waterloo factory in 1927, but most of these were recalled to have design faults rectified. Production started again in the following year with some design modifications and with a name change to General Purpose or GP instead of Model C.

The principal reason for the decision to change the name was the risk of confusing the letters C and D, particularly over a difficult phone line. Another factor was the need for a name which would have a similar meaning to the Farmall name chosen by International. Rumeley also announced a new model in 1928, choosing the name DoAll to emphasise its versatility.

A power operated implement lift was available for the GP. It was not, as is sometimes claimed, the first time this feature had appeared on a farm tractor, but the GP was almost certainly the first production tractor with four different power sources–implement lift, belt pulley, power take-off shaft and drawbar.

**General Purpose
no 208369 built in 1929
and owned by Mike
Twiss from Milton,
Ontario**

The tractor was available in a wide range of versions to meet the needs of different farming systems. These included the wide tread or GPWT which arrived in 1929 and was also the first production John Deere with tricycle front wheels. Early versions of the GPWT had the steering linkage at the side, but this was replaced in 1932 by an overhead steer version with a tapering shape for the hood to give improved visibility from the driving seat.

The GPO or orchard model was added to the General Purpose line in 1931, equipped with side fenders which covered more than half of the rear wheel diameter. Some of the GPO tractors were turned into tracklayers at the Lindeman works in Yakima, Washington, which was the start of a long term association between Lindeman and John Deere. The GPO was the first of a series of orchard models which are among the most interesting and distinctively styled of all the two-cylinder tractors produced by John Deere.

Left

Front view of the 1929/30 GP no 220919 from the Budd brothers' collection . . .

Another of the GP versions was equipped with a 68in tread width to meet the needs of potato growers. Known as the GP Series P, this was added to the range in 1930 and was popular in the potato fields of Maine.

Early versions of the GP were given a 10–20 power rating, meaning 10hp at the drawbar and 20 at the belt pulley, but these figures were easily exceeded in the Nebraska tests. The power output was increased for GPs made after 1930 when the cylinder bore was increased from 5.75 to 6in while the stroke remained at 6in. The C and GP tractors were the first John Deere tractors to be equipped with a three-speed gearbox, a feature which was not available on the Model D until 1935.

Production of GPWT tractors ended in 1933, but the standard and orchard models were still available until 1935.

Below

. . . and the controls

Above
Rear wheel from a 1929 GP

Left
**GP no 220919 in the
Budd collection**

Overleaf
Left **Final drive on a 1929 GP tricycle no 204072 owned by the
Gould family from Michigan**

Right **Front wheels of 204072**

Above
Engine detail . . .

Left
**and single hole main
frame of 204072**

Right
**Engine detail of Chris
Cobler's 1930 GP
Wide-Tread series P**

Above
Protective rear fenders on 1931 GP Orchard model no 15110 owned by Chuck Shuros from Iowa

Left
Engine detail from the GP Orchard

4

New Models for the 1930s

WHILE the GP was approaching the end of its commercial life, John Deere engineers were developing its replacement which was called the Model A. It arrived on the market in 1934 and became one of the most popular of all the John Deere two-cylinder models.

Compared to the GP, the new Model A offered more power, a better view from the driver's seat and a four-speed gearbox instead of the three speeds on the GP. It was equipped with independent rear wheel braking and there was also a splined rear

**Rear wheels
on a 1936 AN**

A rare ANH on 40in rear tyres, made in 1938 and now in the Budd family collection at Ingersoll, Ontario

axle to make it easier to alter the wheel settings. Another major advance was the optional hydraulically operated powered implement lift to replace the mechanical version developed for the John Deere.

The Model A was the world's first production tractor with a hydraulic lift. The new lift was faster than the previous mechanical version, and it also provided a 'cushioning' effect to reduce the risk of damage when an implement was lowered.

Another rare tractor from the Budd collection is this 1938 AWH

Overleaf
The steel wheel A was made in 1934 and the A on rubbers is a 1935 model – both are in the Budd collection

One of the features of the Model A design was the high clearance under the frame and axles, and it was also the first John Deere tractor designed with a one-piece transmission case. Additional clearance for row-crop work was provided by the ANH and AWH high crop versions which were added to the range in 1935 with larger diameter wheels to give more height. The 1950 AH High Crop version took high clearance a stage further with 32in of space under the axles to meet the needs of farmers with special crops such as sugar cane.

The high clearance models were just one example of the wide range of versions of the A which the Deere and Co engineering department developed to help farmers with special crop requirements. The first orchard version, the AO, arrived at the end of 1935 with a more streamlined fender design, and the exhaust and intake stacks no longer extended above the hood.

This 1938 AW is also in the Budd collection

46

AWH photographed by the John Deere publicity department

In 1936 another orchard version was introduced with extra cladding to brush aside fruit tree branches. There was a pointed extension in front of the radiator grille, and a hinged shield covered the steering wheel. It was known as the AOS version, and the disguise was so effective that it was not obvious that the tractor was based on an AR.

Model A production started with a row-crop version with twin front wheels, and the AR fixed tread version was added to the range in 1935. The following year brought the industrial models, usually finished in a distinctive yellow paint colour, and customers were also offered the AN version with a narrow front end, plus various versions with extra-wide wheel settings.

47

An uprated version was launched at the end of 1940. The engine power was boosted by adding an extra 0.25in to the stroke to increase the cubic capacity, and the four-speed transmission was replaced by a three-speed box with high and low ratios to provide six forward gears. This provided an opportunity to increase the maximum forward speed from 6.25 to 11 mph and to give a lower first gear ratio. An electric kit with battery starting and lights was included in the A specification from 1947.

A further addition to the John Deere range arrived at the end of 1935 when the first of the smaller B series tractors left the production line. These were launched with power ratings of 9.28hp on the drawbar and 14.25 on the belt, or about half the power of the first production versions of the Model A.

This is AOS no 1396 built in 1937 and now in the Budd collection

48

A hinged shield gives access to the filler cap on the AOS

1943 AR on rubbers from Aleck Smith's collection at St Marys, Ontario

Many of the developments introduced for the Model A were shared by the B, and these included B equivalents for most of the special A versions. The A and the B were also the models chosen to launch John Deere's new styling in 1938, and they were followed by styled versions of other models in the range.

During the first 40 or more years of tractor development farmers were interested in performance and price rather than appearance, and styling barely registered in the list of design priorities. American automobile manufacturers had already shown that smoother, curving lines could boost sales, and tractor manufacturers adopted the same idea as the competition to win customers intensified during the early 1930s. The new John Deere styling was the work of Henry Dreyfuss, who was employed by the company to replace the familiar angular appearance with more up-to-date lines.

The result was the neat, distinctive lines which first appeared on the A and B tractors and can be traced through more than 20 years until the new generation of tractors was launched in 1961. Deere tractors built before the new look are often referred to as 'unstyled' to distinguish them from the 'styled' tractors with the Dreyfuss treatment.

The two-cylinder engine for the B started with 4.25in bore and 5.25in stroke and the rated engine speed was 1150 rpm. The Model B thus had the distinction of being the first John Deere production tractor with more than 1000 rpm engine speed. The engine size was increased by 0.25in each way to 4.5 × 5.5in in 1938, and there was another increase to 4.69in bore in 1947.

A significant development during the commercial life of the B was the change to an extended frame. This happened in 1937,

54

making it easier to use mid-mounted equipment on the B. A transmission with six forward gears was standard equipment from 1941, and an electric kit was available from 1947.

Orchard versions of the B were first available in 1935, and this led to the famous Lindeman crawler version of the BO. The Lindeman brothers had already converted a small number of GP orchard tractors from wheels to tracks, and the Model B was given the same treatment. Tracklaying versions were used mainly by fruit growers with orchards on sloping land where the extra grip of crawler tracks helped to improve stability. The small size of the B was ideal for orchard work, and Deere shipped about 2000 tractors to the factory between 1939 and 1947 to be fitted with tracks by the Lindeman company. Although most of these tractors were used in orchards, some were equipped with a front blade with hydraulic controls for site work.

Unstyled 1938 Model B Industrial owned by Dave Wickman from Iowa

Front axle of the BI

**1938 BNH no 55195
owned by Bruce
Johnson from Illinois**

After a long association between the two companies, Lindeman was taken over by Deere in 1947, and the factory at Yakima, Washington continued as the specialist production unit for John Deere crawler tractors.

Deere invested heavily in research and development work during the 1930s, and one of the results was the G which arrived in unstyled form in 1938. The new model should have been called the Model F, and F was used as the prefix for part numbers, but Don Macmillan and Russell Jones in *John Deere Tractors and Equipment* say the letter G was chosen instead to avoid confusion with the model F tractor in the International Harvester range.

With an output of 31.44hp on the belt, the G was the most powerful row-crop tractor in the John Deere range. Engine dimensions were 6.13in bore and 7in stroke and the rated engine speed

The main lights on the styled B are supported by the steering column

was 975 rpm. All unstyled Model G tractors were equipped with a four-speed gearbox, but this was replaced by a six-speed transmission in 1941 when the styled version was introduced. The standard specification included a power take-off, and electric starting was standard from 1947.

During the first year of G production the radiator capacity was increased to reduce the risk of overheating, and a wartime GM version was introduced. There were also several specials including a GH model which was equivalent to the High Clear AH tractor.

1942 styled BN pictured at the 1991 Strumpshaw rally in Norfolk

One of the most interesting of the 1930s newcomers was the Model L, which was joined in 1941 by the LA with a slightly bigger engine. They were classified as utility tractors, or what would now be called compacts, and were designed for small farms, large gardens and for working on golf courses and other turf areas.

Stage one in the L/LA development programme was an experimental 8hp tractor known as the Model Y. It was powered by a two-cylinder Nova engine with vertical cylinders. The Y was followed in 1937 by a small production batch of the Model 62 tractor, which was based on the Y prototype and carried a distinctive JD symbol on the rear axle casting and the casting under the radiator. The 62 was built in the John Deere Wagon Works at Moline and was powered by a vertical two-cylinder engine made at the Hercules factory to a John Deere design.

Design features of the 62 and its successors included an offset steering wheel position and a three-speed gearbox. Hand starting was standard, but an electric start kit was available for production models.

By the end of 1937 the 62 tractor had been replaced by the Model L, using the engine developed for the 62. The L was described as a one-plow tractor, capable of turning a 12in furrow in most soil conditions, and in its Nebraska test the power output was about 7hp at the drawbar. An industrial version, called the LI, was added to the range in 1939, and in the same year the L was styled to produce the very distinctive appearance which continued until L production ended in 1946.

An additional model joined the utility tractor range in 1940 when the first of the LA tractors left the Moline factory. The big difference was the increased power, with rated outputs of 10.46hp at the drawbar and 12.93 on the belt pulley, and the LA was able to pull a 16in plow. The extra power was achieved by increasing the 3.25in bore of the L engine to 3.5in for the LA, with 4in stroke for both, and the governed engine speed went up from 1550 for the L to 1850 for the LA.

LA rear wheels were bigger than those on the L, and the 43in standard tread setting was changed to 38, 48 or 54in by moving spacers and turning the wheels. A rear axle with adjustable settings was included in the options list.

When production of the LA ended in 1946, the total number of tractors leaving the factory had reached 13,474 over the six-year period.

1938 unstyled Model G rowcrop from the Budd collection

Rear view of the 1938 G showing the spring suspension under the seat

The 'gate' for the four-speed gear shift on an unstyled G

The last of the 1930s newcomers was the Model H, which began leaving the production line during the last months of 1938. It was another small utility model with approximately the same power output as the LA which followed two years later.

The H engine was a typical John Deere two-cylinder horizontal design. The bore and stroke were 3.56in and 5in respectively, the governed engine speed was 1400rpm, and the power output figures from Nebraska were 12.97hp on the belt and 9.68hp at the drawbar. All the H tractors were styled, and production started with the dual front wheel version. Three other versions followed: the HN with a single front wheel, a high clearance HNH version plus the high clearance HWH with a wide front axle. HNH and HWH tractors were produced for only nine months in 1941/2.

All H tractors were equipped with a three-speed gearbox and the clutch was hand operated. Production ended in 1947 when the John Deere M arrived on the market.

Left
**Front view of the
unstyled G**

Right
Hi-Crop version of the
Model G

Right
Harold Mansfield
brought this 1938
Model L no 622122
from Ohio to the 1993
Two-Cylinder Center
opening at Grundy
Center, Iowa

Left
Rear view of Steve
Barron's G

Above left
1945 LI no 51304 owned by Karl List from Wisconsin photographed in its roped-off enclosure at the Two-Cylinder Center opening

Below left
1946 LA complete with plough owned by the Riseborough family from Norfolk

Right
Two-cylinder vertical engine from the 1946 LA

Below
The single-furrow plough on the Riseborough LA

Above
**1943 HN exported to
England and owned by
I J Smith of King's
Lynn, Norfolk**

Engine detail of the HN

**Narrow rowcrop wheel
on the 1943 HN**

5

The Last of the Letters

1949 M owned by Mike Twiss from Milton, Ontario

APART from specification changes and some additional versions of existing models, tractor development was restricted during the World War II period, but the policy of introducing new models was quickly resumed after the war and the first result was the Model M which arrived in 1947.

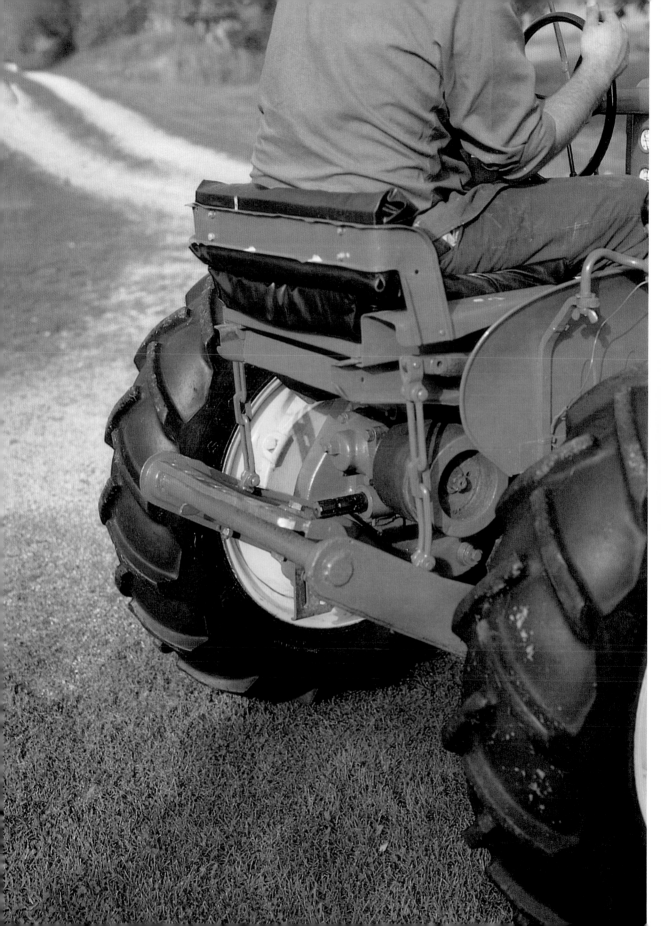

The M was the first tractor to be built at the new Deere factory in Dubuque, Iowa, and it was designed as a more powerful replacement for the L, LA and H utility models. A special feature of the M design was the engine, which continued the John Deere tradition of two cylinders but placed them upright like the L and LA engines instead of the more usual horizontal design. Another unusual feature was the 'square' 4in by 4in cylinder size, the first John Deere engine with equal bore and stroke measurements since the 1930 version of the GP engine.

Power output was rated at 14.39hp at the drawbar and 18.21 on the belt with 1650rpm governed engine speed. The list of standard equipment on the M included a p-t-o, electric starting and a four-speed transmission, and it was also the first tractor fitted with the Touch-O-Matic hydraulic system.

Touch-O-Matic was the name the Deere marketing department chose for the new hydraulic control system for mounted equipment. The name was in line with Roll-O-Matic, the term used for an improved steering system introduced by Deere at about the same time. A later version of Touch-O-Matic provided two levers allowing the driver to use hydraulic power to adjust the level of mounted equipment.

The updated version, known as Dual Touch-O-Matic, first appeared in 1949 when the MT version of the M was announced with the choice of a single wheel, twin front wheels with Roll-O-Matic, or a front axle with adjustable width settings.

Also new for 1949 were the MI industrial version and a crawler model known as the MC. The MC was assembled at the former Lindeman factory in Washington, using skid units supplied from Dubuque, replacing the earlier Lindeman conversions as the first 'all John Deere' tracklayer. Tracks were available in 10, 12 and 14in widths and the tread spacing was adjustable from 36 to 42in.

Top right **Side view of a 1951 MT**
Bottom right **Side view of a 1948 M with lights and downswept exhaust**

Below **Side view of 1950 MI 10001**

The M was a popular and successful addition to the range, but an even more important arrival was the Model R announced in 1948 and available in dealers' showrooms the following year. The R was the last of the long line of Waterloo Boy and John Deere tractors to be known by a letter instead of a model number, and it also introduced a new look for John Deere based on a modified version of the Dreyfuss styling with a redesigned front grille. It was also the most powerful tractor the company had produced at that time and, more significantly, the R was also the first John Deere tractor with a diesel engine.

Development work on a John Deere diesel had continued during the war years, and the result was a two-cylinder horizontal

1954 Model R no 19757 owned by Joel Armistead of Kentucky. Its first owner was a rice grower from Arkansas

design with 5.75in by 8in bore and stroke. This was the engine used for the Model R, capable of delivering up to 51 brake hp and equipped with a petrol starting engine.

Some of the diesel engines available in the mid 1940s had earned a reputation for being difficult to start, particularly in cold weather when the high compression ratio meant they could quickly exhaust the battery of an electric starter motor. Deere, like other American tractor companies offering diesel power at that time, solved the problem by providing a small petrol engine to crank the diesel. As well as reducing the risk of taking all the power out of the battery, the petrol engine also provided waste heat to warm the diesel and make it easier to start.

Driving wheel with rice lugs on Joel Armistead's R

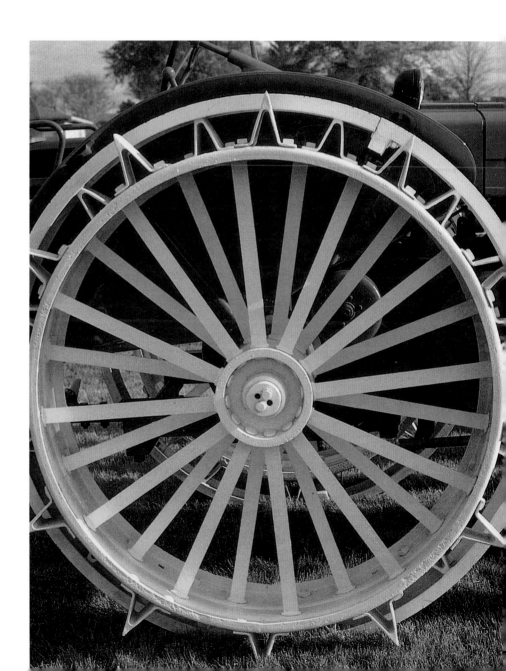

The starting engine was also a two-cylinder design, but with the cylinders horizontally opposed. It was equipped with its own electric starter motor powered from the battery, but with hand starting available as a back-up.

Model R tractors were developed as the replacement for the D, which was ending the longest production run of any John Deere model. The biggest advantage of the switch to diesel power was improved fuel economy, which was emphasised when the R emerged from its Nebraska test with a new record for fuel efficiency.

Features offered on the R for the first time in the John Deere range included a five-speed gearbox, and the list of options included the first steel cab available for a Deere tractor and a p-t-o which could be operated independently of the transmission.

The Model R achieved a considerable success during its six-year production run, firmly establishing Deere as one of the most technically advanced companies in the increasingly popular diesel sector of the market.

1953 Model R without hydraulics or p-t-o, part of the Budd collection

Engine close up of the Budd's Model R

Model R photographed by the John Deere publicity department

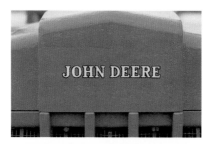

6

New Tractors with Driver Appeal

**1954 40 Utility owned
by John Bosomworth
of Clifford, Ontario**

THE switch from letters to numbers for identifying John Deere tractors came in 1952 with the first instalment of the biggest new model launch Deere had so far made. The A, B, D, G, M and R models were all phased out during the period from 1952 to 1954, to be replaced by the first batch of new numbered models.

The new arrivals were the 40, 50, 60, 70 and 80 models, all featuring front-end styling based on the new look developed for the Model R. The list of important technical developments introduced on the new range included independent carburation for each cylinder, a 'live' p-t-o with independent control, a more efficient cooling system with a belt driven water pump; and an improved version of the Powr-Trol hydraulic system was provided for the 50 and 60 tractors. Another significant development was the increasing importance of driver comfort and convenience with improved accessibility for some of the important controls, more comfortable seats and easier steering.

M series tractors were replaced by the 40 which, like its predecessor, was made at Dubuque and was powered by a twin cylinder 4in × 4in vertical gasoline engine. The power output for the 40 was raised by about 4.5hp by increasing the governed engine speed to 1850rpm, a 200rpm increase, and the standard, tricycle and crawler versions of the M all had their equivalents in the 40 series.

40 C production started in 1953

40 S made in 1953 with mid-mounted 410-S cultivator shown at the 1993 Two-Cylinder Center Grand Opening by Gerald Esslinger of Bayard, Iowa

Aleck Smith of St Marys, Ontario owns this 50 made in 1952

Above
Two 80s joined to give
four-wheel drive and
about 130 hp,
photographed near
Lacombe, Alberta,
Canada

Left top
1956 lpg 60 Orchard
model no 6064003

Left bottom
Fuel tank controls of
6064003

Overleaf
320 S no 321939 (left)
and 320 Utility
no 321932 (right), both
in the collection of
John Bosomworth of
Clifford, Ontario

A new development on the numbered tractors was the addition of power steering as an optional extra—further evidence of the growing importance of the person in the driver's seat. The system, which was hydraulically operated, arrived in 1954 on the 50, 60 and 70 models and was also offered on the 80 which arrived in 1955 and was available only with diesel power.

The 50, 60 and 70 series tractors were all available with a liquefied petroleum gas or lpg engine option as well as gas and 'all fuel'. The lpg versions were modified with a large cylindrical tank instead of a standard fuel tank, giving the tractors a distinctive appearance. Lpg was a popular option for farmers near oilfields where the gas was available at a price low enough to compensate for the power loss.

Another power option for the 70 tractor was a new diesel engine with 6.12in bore and 6.37in stroke. This was considerably bigger than the power unit in the Model R, although the output of the two tractors was almost the same. The diesel version of the 70 arrived in 1954, and the same engine appeared a year later in the new 80 model with the output boosted to 57.49 bhp, making it the most

Text continued on page 93

Vertical engine in the
Dubuque built 320

John Bosomworth's
420 W made in 1957

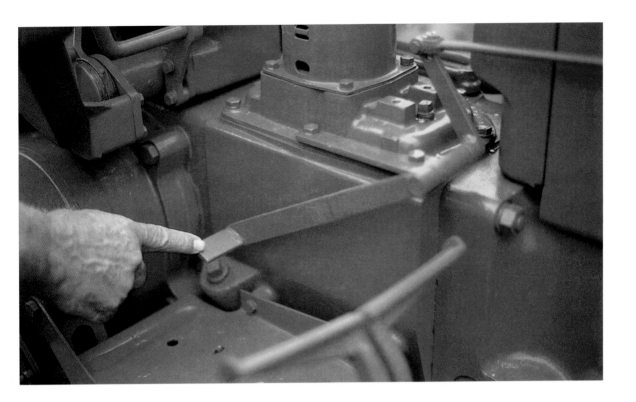

1958 model 420 T showing the foot operated throttle control

powerful production tractor the company had built. It was also economical, establishing a new Nebraska fuel efficiency record and re-enforcing Deere's reputation at the forefront of diesel engine design.

Derivatives of the main models in the numbered series included Hi-Crop versions of the 40, 60 and 70 tractors, and there was also an orchard version of the 60.

Most of the new models in the numbered series arrived on the market during 1952–53, but they were all replaced in 1956 when another numbered series was announced.

The new arrivals were the 20 series, starting with the 320 and 420 utility models from Dubuque which replaced the previous 40 model, and going through to the 64hp 820 diesel. The most obvious difference between the 20 series and the previous models was the new eye-catching green and yellow colour scheme, but there were also some significant mechanical differences, including an improved hydraulic control system for mounted equipment.

Custom Powr-Trol was the name chosen for the 20 series hydraulics. The important advantages claimed for the system included more accurate control of implement working depth, and there was a special control to ensure that the linkage returned the implement to the same depth after being lifted for a headland turn. The system also provided automatic weight transfer from the implement to improve traction.

Some of the 20 series engines were inherited from the previous model range, but with detail improvements to the combustion chamber design, and these included the 4in × 4in gas engine for the 320, the 4.69in × 5.5in engine for the 520, and the diesel engines from the 70 and 80 tractors appeared again in the new 720 and 820 diesel tractors. There were also some basically new power units introduced for the 20 series, including an 'over-square' 420 engine with 4.25in bore and a smaller 4in stroke.

Driver comfort moved a further step forwards with power steering now included in the standard specification of all but the 320 and 420 tractors. The suspension for the new Float-Ride seat could be adjusted to match the weight of the driver, and there was more space for the driver's feet.

The 420 tractor was available in the widest range of options including two 420C crawler models, one with four-roller tracks and the other with five rollers. Other versions available in the 20 series included an attractive orchard or grove 620 plus some Hi-Crop models.

520 rowcrop from John Bosomworth's collection

620 showing the two-colour paint finish introduced on the 20 series

Beautifully restored 620 Orchard with KBA discs owned by Shiela & Don Morden of Moorefield, Ontario

Overleaf
**1958 rowcrop 720 (left)
with a 1956 standard
version pictured at the
1992 Steam Era event
at Milton, Ontario**

The 820, like its predecessor the 80, was equipped with a diesel engine as standard equipment. The engine was uprated in 1957 with improved fuel injection and a different piston design to gain an 11 per cent increase in power output to 75.60hp on the belt. Several versions of the 820 were available, including an industrial model and a Rice Special with extra protection for the brakes and rear axle bearings, and with special tyres.

After only two years on the market the 20 series gave way to the mechanically similar 30 range. With the mechanical specifications generally similar for the two series, the important improvements were once more aimed at driver comfort, with operator safety again emerging as a significant factor.

Text continued on page 102

**Rear view of the
Mordens' 620 Orchard**

Right **This 720 rowcrop owned by Willis Richardson of Arthur, Ontario is one of the last to be built and has 30 series fenders and electric starting**

Below **1956 standard 720 no 7204652**

New safety considerations on the 30 series included handholds and a step to make mounting and dismounting less hazardous. Lighting equipment was improved for increased safety as well as to make night working easier, there were detail improvements to the seat design and the steering wheel position, and the instrument layout was improved and offered additional information.

One of the few important mechanical developments was an electric start option for the 730 and 830 diesel tractors, an indication that the latest diesels were easier to start and evidence that the petrol starting engine was becoming obsolescent.

The increasing importance of diesel power for tractors was emphasised in 1959 when the last new two-cylinder model was added to the John Deere range. It was based on the 430, but the model number was 435, to set it apart from the 30 series, and the power unit was a General Motors two-cylinder diesel engine with 3.87in × 4.5in bore and stroke.

Auxiliary fuel tank and starter engine on the 720

Diesel power for the 720

Nebraska had recently changed its test procedures to put more emphasis on p-t-o performance instead of belt power. The 435 was the first John Deere tractor to be tested under the new system, and it was also the first to be offered with the choice of 540 or 1000 rpm p-t-o speeds. The output recorded in the Nebraska p-t-o test was 32.91, with 28.41hp at the drawbar.

Only about 2500 of the 435 tractors were built during the two-year production run, but it was significant as the first John Deere model to provide the benefits of diesel power for smaller farms and equipment.

A new tractor range announced in 1961 brought the reign of the two-cylinder engine to an end at John Deere. It was probably the most successful engine series the tractor industry had produced, and when John Deere announced the 'New Generation of Power' in 1960, with more to come in 1963/64, the emphasis was on four-cylinder engines.

1957 Model 820 photographed at the 1992 Steam Era event, Milton, Ontario

30 Series collection owned by the Mordens of Moorefield, Ontario

1959 330 S no 330567 from the Mordens' collection

It was a clean break. Apart from the end of the two-cylinder engine series, the new tractors also introduced totally new styling which left no remaining traces from the Dreyfuss lines of 1938, new transmissions including a powershift, and new hydraulics. In fact 95 per cent of all the components were new.

What remains today are thousands of two-cylinder models which have been preserved by enthusiasts as a permanent reminder of some of the most famous and successful tractors from the past.

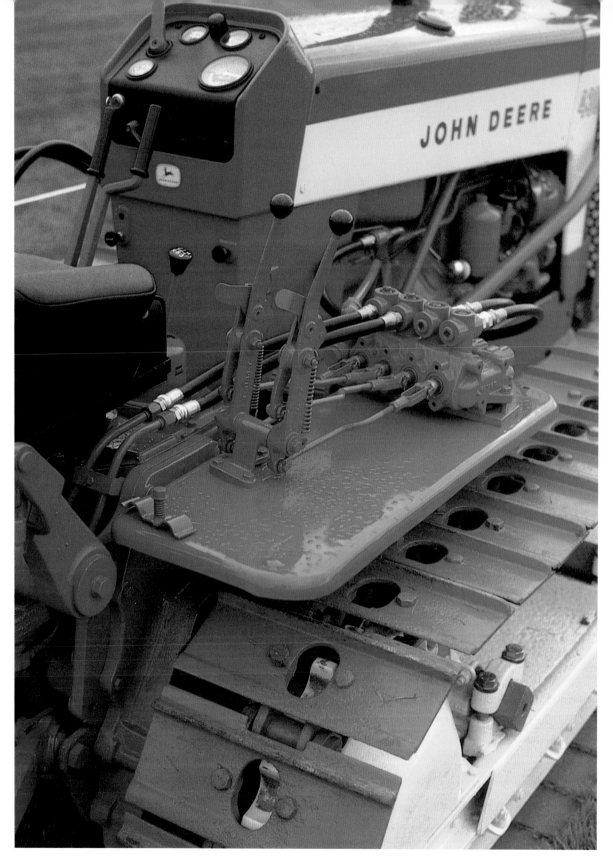

Above
Rick Mabary of Emerson, Iowa brought this 730 Standard lpg to the 1993 Two-Cylinder Center opening

Right
lpg tank on a 530 owned by Lowell Kroneman of Iowa

Left
430 C no 143042 showing the hydraulic controls and snowshoe tracks

Above
830 Rice Special owned
by Kenneth Peterman
from Iowa

Right
2OR 18-26 rear tyre on
the 830 Rice Special

Left
Steps for the driver to
climb aboard on Rick
Mabary's 730 lpg

John Bosomworth's
435 powered by a GM
diesel engine

Appendix A

John Deere at Nebraska

WHEN Deere and Co was developing the All-Wheel Drive or Dain tractor during World War I they spent four years improving and testing prototype versions before approval was given for a limited production run.

Other manufacturers were not always as careful. Because of the great wartime demand for tractor power, some of the new models being rushed on to the market were badly designed and were sold with little or no service or parts support. Farmers who had previously worked with horses or steam engines often knew little or nothing about tractor design, and many of them were disappointed by the poor reliability and inadequate performance of their first tractor. The situation was similar in Canada, and the authors of *The Adoption of the Gasoline Tractor in Western Canada*, published in 1980, quote a survey by an Alberta Department of Agriculture official in 1921 which concluded that 80 per cent of Alberta farmers who owned tractors regretted the investment.

The situation became so serious that farmers and some of the more responsible manufacturers demanded an independent test programme which would provide information to make it easier for customers to choose a better tractor. It was an idea that attracted Wilmot F Crozier, a farmer from Polk County, Nebraska who had owned several badly designed tractors and was well aware of the problems they could cause.

He was also involved in politics, and he introduced the bill in the state legislature which became the Nebraska Tractor Test Law. This made it illegal to sell a tractor in Nebraska on the basis of false claims and without a permit.

The permit could not be issued until a tractor of the same model had completed the official test programme to prove that the performance matched the manufacturer's claims. The manufacturer was also required to show that adequate parts and service back-up were available for the tractor.

The bill was passed in 1919 and the responsibility for carrying out the test programme was given to the Agricultural Engineering Department of the University of Nebraska. Preparations were completed quickly and the engineers were ready to begin their first test towards the end of the year with a Twin City 12-20 tractor.

This test was abandoned because of heavy snow, and when the programme was resumed in the following spring a Waterloo Boy Model N entered by Deere became the first tractor to complete a Nebraska test successfully. The Twin City tractor was tested later and is listed as number 19.

Since then the Nebraska test reports have continued to provide an internationally recognised measurement of tractor performance, and they are also a valuable record of the improvements in efficiency achieved by manufacturers during more than 70 years of tractor development. One example is fuel efficiency, which in the Nebraska tests has been measured in hp hours per gallon or the amount of horsepower available for one hour from one gallon (US) of fuel. The Waterloo Boy in the first Nebraska test developed the equivalent of 6.83hp for one hour from one gallon of fuel in the belt test, but when the 720 diesel was tested 36 years later this had risen to 17.97hp hrs.

As well as showing the general improvement in fuel efficiency, the John Deere tractor tests in the following table are also a good example of the differences in efficiency for the various fuels used by the tractors over the period covered by two cylinder production. The 17.97 hp hrs/gal achieved by the 720 diesel in test number 594 compares with only 9.27 hp for the lpg version of the same model in test 593.

Staff at the university completed 65 tractor tests during 1920, and the results for the year show wide variations between the performance figures achieved by different tractors.

An example is water loss from the cooling system, which was clearly a major problem with some of the tractors tested. The Avery 25-50 evaporated—or perhaps leaked—25 gallons during the maximum load belt test lasting one hour, and the loss from the 40-80 tractor from the same manufacturer was 18.75 gallons in the same test. The 0.83 gallons lost by the Waterloo Boy was better than average and a considerable improvement over rivals such as the International Harvester 10-20 Titan which lost 7 gallons.

Figures for wheelslip in the ten hour test of drawbar performance showed lightweight tractors were at a disadvantage for jobs such as pulling a plough. The Model F Fordson weighing 2710lb had the worst wheelslip figure of all the agricultural tractors tested in 1920, the Waterloo Boy weighed 6183lb and returned 11.84 per cent slip, and the best figure for a wheeled tractor came from the 25,510lb Twin City 40-65 with 2.00 per cent. The best grip in the 1920 tests was the 0.60 per cent from a Holt T-16 tracklayer.

Early tests also recorded the quantity of lubricating oil tractors used during their complete programme of tests, and the results show some 1920 models needed frequent topping-up. Oil consumption by the Waterloo Boy was among the lowest 20 per cent at about 3.5 gallons in 44 hours of tests, the Titan 10-20 used

8 gallons in 30 hours and the figure for a Case 15-27 was 11.5 gallons. The most remarkable figures were from an Aultman-Taylor 30-60 with 17.5 gallons used in 44 hours, and 29 gallons disappeared during the 53 hour test programme for the Case 22-40.

After officially starting the Nebraska test series, tractors from the John Deere range have continued to figure prominently in the tests, and abbreviated results for some of the tractors are in the following table.

Model	fuel	number	year	hp	hp hrs per gal	hp	pull lb	speed mph	slip %
		Belt Test Max load—one hour test				Drawbar Test Rated load—10 hours			
WB N	Ker	1	20	25.97	6.83	12.10	1982	2.29	11.84
D	Ker	102	24	30.40	9.03	16.75	1786	3.52	6.58
D	Ker	146	27	36.98	9.29	15.58	1577	3.66	5.85
GP	Ker	153	28	24.97	9.18	10.20	1078	3.55	1.01
GP	dis	190	31	25.36	9.50	15.34	1702	3.38	2.23
A	dis	222	34	24.71	10.10	16.31	1839	3.33	3.30
B	dis	232	34	15.07	10.19	9.28	1023	6.39	0.77
D	dis	236	35	40.11	10.14	24.64	2357	3.92	3.19
G	dis	295	37	34.09	10.59	20.75	2252	3.46	3.70
B	dis	305	38	18.31	10.56	10.84	1228	3.31	0.66
B	dis	305	38	18.31	10.56	13.12	1195	4.12	3.70
H	dis	312	38	14.22	11.95	9.77	1065	3.44	4.98
L	gas	313	38	10.42	9.82	7.06	688	3.85	4.98
A	dis	335	39	28.93	11.30	20.48	1815	4.23	2.14
D	dis	350	40	40.24	10.14	30.77	2907	3.97	7.30
B	dis	366	40	19.69	11.62	17.13	2095	3.07	8.89
LA	gas	373	41	14.34	10.47	10.62	1091	3.65	4.49
AR	dis	378	41	28.71	11.12	20.66	1861	4.16	4.60
B	gas	380	47	25.79	11.79	19.04	1666	4.29	4.78
B	dis	381	47	22.17	11.49	16.82	1908	3.31	5.83
G	dis	383	47	36.03	10.74	27.08	2313	4.39	5.14
A	gas	384	47	35.81	11.44	26.70	2441	4.10	6.12
M	gas	387	47	19.49	11.12	14.65	1279	4.30	6.80
R	dsl	406	49	48.58	17.35	34.45	3140	4.11	4.70
MT	gas	423	49	19.80	11.16	14.15	1118	4.75	3.77
R	gas	429	49	36.13	11.74	26.24	2279	4.32	4.71
MC	gas	448	50	20.12	11.36	13.88	1833	2.84	1.92
60	gas	472	52	38.58	11.61	28.04	2402	4.38	6.18
50	gas	486	52	28.85	11.82	20.89	1747	4.48	5.74
60	t/f	490	53	31.09	11.08	22.69	1918	4.44	4.27
70	gas	493	53	45.88	11.92	33.61	2771	4.55	5.35
40	gas	503	53	23.51	11.16	17.43	1475	4.43	5.67
40S	gas	504	53	23.21	11.19	17.05	1364	4.69	6.22
40C	gas	505	53	23.64	11.61	15.08	1938	2.92	1.57

		Belt Test Max load—one hour test				Drawbar Test Rated load—10 hours			
Model	fuel	number	year	hp	hp hrs per gal	hp	pull lb	speed mph	slip %
70	t/f	506	53	40.90	11.06	31.06	2539	4.59	3.93
50	t/f	507	53	23.64	11.09	17.63	1904	3.47	4.68
60	lpg	513	53	38.94	9.16	28.78	2443	4.42	4.47
70	lpg	514	53	48.18	9.57	34.95	2878	4.55	4.99
70	dsl	528	53	50.40	17.74	34.79	2831	4.61	4.33
50	lpg	540	55	30.18	9.13	22.13	1858	4.47	5.12
40S	t/f	546	55	19.13	10.51	14.35	1329	4.05	6.10
80	dsl	567	55	65.33	17.58	46.83	3979	4.41	6.19
520	lpg	590	56	35.75	10.27	26.10	2229	4.39	3.95
620	lpg	591	56	48.13	9.60	34.53	2916	4.44	4.78
520	t/f	592	56	24.75	10.94	18.95	2087	3.41	3.60
720	lpg	593	56	55.48	9.27	39.05	3428	4.27	5.26
720	dsl	594	56	56.66	17.97	40.41	3532	4.29	5.12
520	gas	597	56	36.11	12.74	26.05	2236	4.33	3.74
620	gas	598	56	44.25	12.52	33.59	2834	4.44	4.26
420W	gas	599	56	27.25	11.72	20.82	1689	4.62	4.50
420S	t/f	600	56	22.73	9.98	17.07	1482	4.32	5.24
420C	gas	601	56	27.39	11.90	18.65	2407	2.91	2.09
620	t/f	604	56	32.87	11.25	24.63	2065	4.47	3.36
720	gas	605	56	55.11	12.21	40.37	3499	4.33	3.97
720	t/f	606	56	42.38	10.89	31.59	2708	4.38	3.03
820	dsl	632	57	72.82	17.28	53.16	4553	4.38	5.76

For the following test the maximum load belt figure was replaced by a more up-to-date test with output measured at the p-t-o, and drawbar pull figures were measured at 75 per cent load in place of the rated load drawbar test.

435	dsl	716	59	32.91	14.35	22.54	1557	5.43	3.00

Notes:

1. All tractors in the first 10 tests in the table were on steel wheels, and the rest on rubber tyres. The exceptions are test numbers 448, 505 and 601, which were tracklayers.

2. Fuels. Ker is kerosene or paraffin, dis is distillate, gas is gasoline or petrol, t/f is tractor fuel or tvo, dsl is diesel and lpg is liquefied petroleum gas.

Appendix B

Waterloo Boy and John Deere Tractor Serial Numbers

(Note: Data supplied by Deere & Co. The years in most cases are production years from August 1st to July 31st of the year stated. Exceptions are marked with an asterisk.)

WATERLOO BOY

Year	L/LA	R	N
1914	1,000		
1915		1,026	
1916		1,401	
1917		3,556	10,020
1918		6,982	10,221
1919		9,056	13,461
1920			18,924
1921			27,026
1922			27,812
1923			28,119
1924			29,520

JOHN DEERE

Year	D	GP Std	GP Wide	GP 0	G/GM
1924	30,401				
1925	31,280				
1926	35,309				
1927	43,410				
1928	54,554	200,111			
1929	71,561	202,566	400,000		
1930	95,367	216,139	400,936		
1931	109,944	224,321	402,741	15,000	
1932	115,477	228,666	404,770	15,226	
1933	115,665	229,051	405,110	15,387	
1934	116,273	229,216		15,412	

(Continued)

JOHN DEERE (*continued*)

Year	D	GP Std	GP Wide	GP 0	G/GM	
1935	119,100	230,515		15,589		
1936	125,430					
1937	130,700					
1938	138,413				1,000	
1939	143,800				7,734	
1940	146,566				9,321	
1941	149,500				10,489	
1942	152,840				12,069	
1943	155,005				12,941	
1944	155,426				13,748	
1945	159,888				13,905	
1946	162,598				16,694	
1947	167,250				20,527	
1948	174,879				28,127	
1949	183,516				34,587	
1950	188,420				40,761	
1951	189,701				47,194	
1952	191,180				56,510	
1953	191,439				63,489	

Year	A	AO/AR	AO styled	B	BO/BR	BO Lind
1934	410,000					
1935	412,869			1,000		
1936	424,025	250,000		12,012	325,000	
1937	442,151	253,521	AO 1000	27,389	326,655	
1938	466,787	255,416	AO 1539	46,175	328,111	
1939	477,000	257,004	AO 1725	60,000	329,000	
1940	488,000	258,045	AO 1801	81,600	330,633	
1941	499,000	260,000		96,000	332,039	
1942	514,127	261,558		126,345	332,427	
1943	523,133	262,243		143,420	332,780	332,901
1944	528,778	263,223		152,862	333,156	333,110
1945	548,352	264,738		173,179	334,219	333,666
1946	555,334	265,870		183,673	335,541	335,361
1947	578,516	267,082		199,744	336,746	336,441
1948	594,433	268,877		209,295		
1949	620,843	270,646		237,346		
1950	648,000	272,985		258,205		
1951	667,390	276,078		276,557		
1952	689,880	279,770		299,175		
1953		282,551				

116

Year	H	62/L	LA	M	MC	MT	R
1937		621,000					
1938		621,079					
1939	1,000	626,265					
1940	10,780	630,160					
1941	23,654	634,191	1,001				
1942	40,995	640,000	5,361				
1943	44,755	640,738	6,029				
1944	47,796	641,038	6,159				
1945	48,392	641,538	9,732				
1946	55,956	641,958	11,529				
1947	60,107			10,001			
1948				13,734			
1949				25,604	10,001	10,001	1,000
1950				35,659	11,630	18,544	3,541
1951				43,525	13,630	26,203	6,368
1952				50,580	16,309	35,845	9,293
1953							15,720
1954							19,485

Year	40 Std	40 Tricycle	40 Hi Crop	40 Spl	40 U	40 U(2-r)	40C
1953	60,001	60,001			60,001		60,001
1954	67,359	72,167	60,001		60,202		63,358
1955	69,474	75,531*	60,060	60,001	63,140	60,001	66,894

Year	50	60	70	80
1952	5,000,001	6,000,001		
1953	5,001,254	6,007,694	7,000,001	
1954	5,016,041	6,027,995	7,005,692	
1955	5,021,977	6,042,500	7,017,501	8,000,001
1956	5,030,600	6,057,650	7,034,950	8,000,755

Year	320	420/420C	520	620	720	820
1956	320,001	80,001	5,200,000	6,200,000	7,200,000	8,000,000
1957	321,220	107,813	5,202,982	6,203,778	7,203,420	8,200,565
1958	325,127*	127,782*	5,209,029	6,215,048	7,217,368	8,203,850

Year	330	430/430C	530	630	730	830
1958	330,001	140,001	5,300,000	6,300,000	7,300,000	8,300,000
1959	330,171	142,671	5,301,671	6,302,749	7,303,761	8,300,727
1960	330,935	158,632	5,307,749	6,314,381	7,322,075	8,305,301
1961					7,328,801	8,306,892

Year	435
1959	435,001
1960	437,655

* 40 Tricycle 1955 - fiscal year Nov 1 to Oct 31
  320, 420/420C 1958 - calendar year.

Index

FARMING PRESS BOOKS & VIDEOS

Below is a sample from the wide range of agricultural and veterinary books and videos published by Farming Press. For more information or for a free illustrated catalogue of all our publications please contact:

**Farming Press Books & Videos, Wharfedale Road
Ipswich IP1 4LG, United Kingdom
Telephone (0473) 241122   Fax (0473) 240501**

VHS COLOUR VIDEOS

The Massey-Ferguson Story
MICHAEL WILLIAMS

Michael Williams takes us from the early days of Wallis and the General Purpose tractor right up to modern high-spec models.

Fordson: the story of a tractor
TOLD BY BOB SYMES

This features the five main Fordson models from 1917 to the 1950s. It combines archive material with new film.

BOOKS

Tractors Since 1889   MICHAEL WILLIAMS

An overview of the main developments in farm tractors from their stationary engine origins to the potential for satellite navigation. Fully illustrated.

Tractors: how they work and what they do   MICHAEL WILLIAMS

For younger readers, a highly illustrated account of tractor principles and development, looking also at larger American and Australian models.

Ford and Fordson Tractors

Massey-Ferguson Tractors
MICHAEL WILLIAMS

Two heavily illustrated guides to the models which made these leading companies great.

A Full Pull   GEOFF ASHCROFT

The sport of tractor pulling described and colourfully illustrated

Farming Press Books & Videos is part of the Morgan-Grampian Farming Press Group, which also publishes a range of farming magazines: *Arable Farming, Dairy Farmer, Farming News, Pig Farming, What's New in Farming.* For a specimen copy of any of these please contact the address above.